Understanding the Elements of the Periodic Table™

ALUMINUM

Heather Hasan

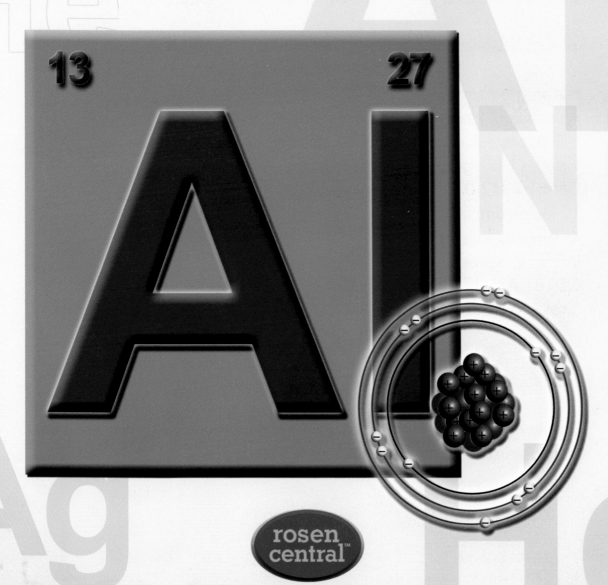

13   27

Al

rosen
central™

The Rosen Publishing Group, Inc., New York

To my mother-in-law, Donna. Thank you for all of your love and support and for making me feel like a welcome part of your family.

Published in 2007 by The Rosen Publishing Group, Inc.
29 East 21st Street, New York, NY 10010

Library of Congress Cataloging-in-Publication Data

Hasan, Heather.
Aluminum / Heather Hasan. — 1st ed.
    p. cm. — (Understanding the elements of the periodic table)
Includes bibliographical references and index.
ISBN 1-4042-0705-8 (library binding)
1. Aluminum. 2. Group 13 elements. 3. Periodic law.
4. Chemical elements.
I. Title. II. Series.
QD181.A4H37 2007
546'.673—dc22

                                                    2005032022

Manufactured in the United States of America

On the cover: Aluminum's square on the periodic table of elements; the atomic structure of an aluminum atom *(inset)*.

Contents

Introduction

Aluminum is an extremely useful metallic element with the symbol Al. It is light and nontoxic, and it does not corrode. It is used domestically for kitchen utensils, aluminum foil, and cans. Aluminum's strength and light weight also make it ideal for the construction of automobiles, railroad cars, airplanes, and spacecraft. As a pure element, aluminum is a fairly soft, bluish white metal. However, in nature, aluminum is never found in its pure state.

While other metallic elements, such as copper (Cu), tin (Sn), and lead (Pb), have been used by man for thousands of years, most scientists and scholars agree that aluminum metal was not discovered until about 200 years ago. However, some scholars have suggested that it may have been discovered by the Romans as long as 2,000 years ago. These scholars believe that aluminum is first mentioned in the writings of Pliny the Elder, an officer in the Roman army and an author. In Book XXXV of his famous encyclopedia, *Historia Naturalis*, published in the first century AD, Pliny describes a silvery metal that bears a strong resemblance to aluminum.

According to Pliny, one day the Roman emperor Tiberius was presented with a strange dinner plate by a goldsmith. The plate was made from a new kind of metal that was both very shiny and extremely light. The goldsmith said that he had made the metal from clay. The goldsmith thought the emperor would be pleased with this news, but he was wrong—the emperor did not like the idea of a valuable metal that could be made from clay. Afraid that the amazing new metal would make all of the gold in his kingdom worthless, he had the goldsmith beheaded.

Nearly 2,000 years after Pliny's account of the strange metal, another emperor, Napoléon III, used aluminum plates at banquets. Aluminum was such a rare and precious metal at that time that only the most important guests were given aluminum plates. Other guests were given plates of pure gold.

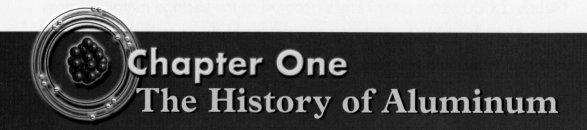

Chapter One
The History of Aluminum

Aluminum is the most abundant metal in the earth's crust. It is also the third most abundant element in the crust (after oxygen [O] and silicon [Si]), making up about 8.2 percent of it. Although it is a very plentiful element, aluminum is never found free in nature. Instead, it is always found combined with other elements.

The most important source of aluminum is a kind of rock called bauxite, which provides us with over 99 percent of the metal. Bauxite is a name for a mixture of minerals (gibbsite, diaspore, and boehmite) that contain aluminum, oxygen, and water. Bauxite is mainly located in tropical and subtropical climates. The largest amount of bauxite is found in Australia, but substantial amounts are also found in Brazil, Guinea, and Jamaica.

The Discovery of Aluminum

The word "aluminum" is derived from the Latin word *alumen*. Alum is a white powder that contains aluminum and other elements, such as sulfur (S). Ancient Roman surgeons applied alum to the wounds of injured soldiers. This closed the open blood vessels and allowed wounds to heal. Alum was also used to make dyes stick to fabric. In the Middle Ages, alum was used to cure, or dry, the skin of dead animals.

Bauxite is a mixture of hydrated aluminum oxides usually containing oxides of iron and silicon in varying quantities. Its color ranges from white or tan to a deep brown or red. The color of the bauxite depends upon its components. Bauxite is soft and claylike or earthy in texture. It is characterized by having small, round, pea-sized lumps. Bauxite can easily be purified and converted into metallic aluminum.

As early as 1787, scientists began to suspect that an unknown metal existed in alum. In the early nineteenth century, a British chemist named Sir Humphry Davy named the unknown metal aluminum. However, scientists did not have a way to extract this metal until 1825. The first scientist to successfully produce tiny amounts of aluminum was a Danish chemist named Hans Christian Ørsted. Two years later, a German chemist by the name of Friedrich Wöhler developed a more efficient way to obtain the metal. By 1845, he was able to produce large enough amounts of aluminum to determine some of its basic properties. Wöhler's method was improved upon in 1854 by a French chemist named Henri-Étienne Sainte-Claire Deville. Deville's method for extracting aluminum allowed for the commercial production of the metal. It also lowered the price of aluminum from $544 per pound ($1,200 per kilogram) in 1852 to $18 per pound ($40 per kilogram) in 1859.

Aluminum on Display

Following French chemist Henri-Étienne Sainte-Claire Deville's development of a commercial method to extract aluminum, the metal was displayed in Paris's Universal Exposition in 1855. At the exposition, a bar of aluminum was exhibited next to the English crown jewels. Aluminum was considered a scientific marvel and was, at that time, more valuable than gold (Au).

Aluminum's display in the Paris exhibition sparked the imagination of jewelry makers, silversmiths, and watchmakers. From 1850 to the late 1870s, aluminum usually made its appearance in the form of small luxury items, such as bracelets, medallions, and opera glasses.

Henri-Étienne Sainte-Claire Deville was born on March 11, 1818, on the island of St. Thomas, West Indies. Educated in Paris at the Collège Rollin, he graduated as a doctor of medicine and doctor of science.

In 1888, Charles Martin Hall, along with financier Alfred E. Hunt, founded the Pittsburg Reduction Company, which is now known as the Aluminum Company of America (ALCOA). Today, ALCOA is the world's leading producer of aluminum. ALCOA materials can be found in wheels (like the one this ALCOA employee in Beloit, Wisconsin, is working on), bike frames, airplanes, motorcycles, and soda cans.

Aluminum remained too expensive for wide commercial use until 1889, when American chemist Charles Martin Hall and French metallurgist Paul-Louis-Toussaint Héroult patented an inexpensive method for the production of pure aluminum. As a result of this process, the cost of aluminum had dropped down to 18¢ per pound (40¢ per kg) by 1914. It wouldn't be long before the aluminum industry became one of the largest and most important metal industries in the world.

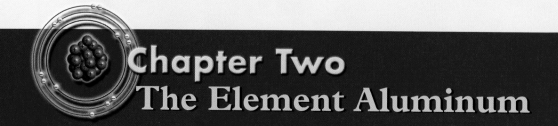

Everything on the earth is made of one or more elements. Each element is made up of only one kind of atom. This means that every atom in the element aluminum is exactly the same.

Atoms are very tiny and cannot be seen by the naked eye. It would take about 500 million atoms, lying side by side, to form a line that is only 1 inch (2.5 centimeters) long. Atoms are made up of even smaller components called subatomic particles. In order to truly understand what makes aluminum unique, we have to take a closer look at these particles.

Subatomic Particles

Atoms are made of three subatomic particles: neutrons, protons, and electrons. Neutrons and protons are clustered together at the center of the atom to form a dense core called the nucleus. Neutrons carry no electrical charge, while protons have a positive electrical charge. This gives the nucleus an overall positive electrical charge. Since aluminum has thirteen protons in its nucleus, its nucleus has a charge of +13.

Electrons are negatively charged particles that are arranged in layers, or shells, around the nucleus of an atom. The negative electrons are attracted to the positive nucleus, and it is this attraction that keeps the electrons spinning rapidly around the nucleus and holds the atom together. The number

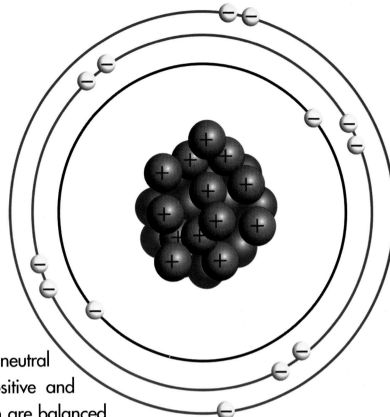

Though the electrons in this diagram appear to be orbiting the nucleus like planets around the sun, electrons actually do not follow any specific path. Instead, electrons form regions of negative electrical charge around the nucleus. These regions are called orbitals and correspond to where an electron is most likely to be found. A group of orbitals is called a shell. Though not entirely accurate, chemists find models such as this very useful.

of protons and electrons in a neutral atom are equal, so the positive and negative charges of the atom are balanced. Therefore, since aluminum has thirteen protons, it also has thirteen electrons.

The Periodic Table

There are more than 100 known elements today. As scientists discovered more and more elements over the years, they realized that they had to be organized somehow. Eventually, the elements were arranged on a big chart, called the periodic table.

The periodic table that we use today is based on the work of a Russian chemist named Dmitry Mendeleyev. He published the first version of the periodic table in 1869, while teaching chemistry at the University of St. Petersburg in Russia. Mendeleyev sought to organize the elements in a way that would make it easier for his students to study and understand them. He arranged the elements in horizontal rows, according to weight, with the lightest element of each row on the left end and the heaviest on the right.

Aluminum Snapshot

Chemical Symbol:	Al
Properties:	Lightweight, ductile metal that is solid at room temperature, conducts heat well, and does not corrode.
Discovered By:	Although it is impossible to say when aluminum was first discovered, a method of extracting the metal was developed in 1825 by Hans Christian Ørsted.
Atomic Number:	13
Atomic Weight (actual):	26.982 atomic mass units (amu)
Protons:	13
Electrons:	13
Neutrons:	14
Density at 68°F (20°C):	2.7 g/cm^3
Melting Point:	1221°F (660°C)
Boiling Point:	4472°F (2467°C)
Commonly Found:	In the earth's crust

Unlike Mendeleyev's chart, the periodic table that we use today lists the elements in order of increasing atomic number. An element's atomic number is equal to the number of protons in its nucleus. By seeing where an element is located on the periodic table, you can predict whether it is a metal, a nonmetal, or a metalloid.

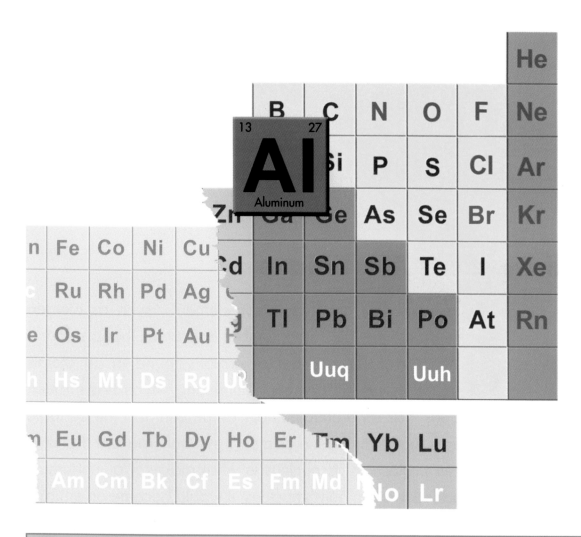

An element's atomic number is listed to the upper left of its symbol. The number found to the upper right of an element's symbol is its atomic weight (the average weight of an atom of the element). The periodic table is a very useful tool. If a person knows the main properties of a group and how the properties of the elements vary within that group, he or she can reasonably predict the properties of any element.

Metals, such as aluminum, are easily recognized by their physical traits. Generally, metals can be polished to be made shiny. Most metals have the ability to be hammered into shapes without breaking, a property called malleability. Metals are usually ductile, meaning that they can be pulled into wires. They also conduct electricity. Since it is a metal, aluminum has all of these physical traits. Substances such as wood, glass, or plastics are classified as nonmetals because they lack the characteristics of metals. Metalloids, or semimetals, have characteristics of both metals and nonmetals.

If you look at the periodic table, you will notice that the elements are divided by a "staircase" line. The metals are found to the left of this line and the nonmetals on the right. Most of the elements bordering the line are metalloids. Though aluminum is one of the elements that border the staircase line, it is the one exception to the rule. Aluminum does not have the characteristics of a metalloid, but rather those of a metal.

All Elements Are Unique

What makes aluminum different from other elements, such as oxygen or silver (Ag)? The difference lies in the number of protons that are found in the nuclei of its atoms. The number of protons found in an atom's nucleus correspond to its atomic number. Aluminum, which has thirteen protons in its nucleus, has an atomic number of thirteen. On the periodic table, this number is found above the element's symbol. The fact that aluminum has thirteen protons in its nucleus is what makes it aluminum. If one proton was added to aluminum's nucleus, it would become an entirely different element. Adding another proton would produce the element silicon, which has fourteen protons in its nucleus. If one of aluminum's protons was removed, it would become magnesium (Mg), which has twelve protons in its nucleus.

Groups and Periods

As you look across the periodic table from left to right, each horizontal row of elements is called a period. Elements are arranged in periods by the number of electron shells that surround the nuclei of their atoms. Aluminum is in period 3, so each of its atoms has three shells of electrons surrounding its nucleus. There are two electrons in aluminum's innermost electron shell, eight in the next, and three in its outermost shell. The electrons in the outermost shell are called valence electrons. These electrons determine how an element acts.

As you read down the periodic table from top to bottom, each vertical column of elements is called a group. All of the elements in a group have the same number of electrons in their outermost electron shell. Aluminum is in group IIIA. The other elements found in group IIIA are boron (B), gallium (Ga), indium (In), and thallium (Tl). Each of these elements has three electrons in its outermost electron shell. However, they have very little else in common with one another and differ widely in the way that they react with other chemicals.

The group IIIA elements become more metallic in nature when moving down the column. Although they are

Each column of the periodic table lists elements that share similar chemical properties. These properties depend upon the arrangement of electrons in the atom. The elements in a group have the same number of electrons in their valence shells. All the elements in Group IIIA, including aluminum, have three electrons in their outermost shells.

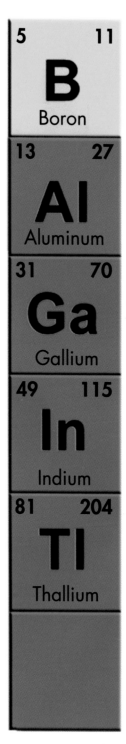

5	11
B	
Boron	

13	27
Al	
Aluminum	

31	70
Ga	
Gallium	

49	115
In	
Indium	

81	204
Tl	
Thallium	

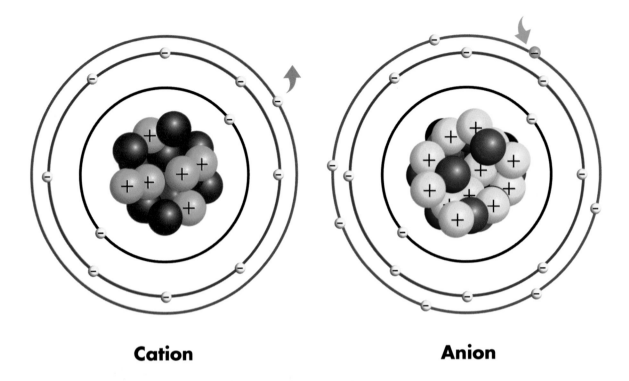

Cation   **Anion**

An ion is formed when an atom loses or gains one or more negatively charged electrons. When an atom loses one or more electrons, it becomes a positively charged ion, called a cation. When an atom gains one or more electrons, it becomes a negatively charged ion, called an anion. Here, a sodium atom loses an electron to form a sodium cation (Na+), and a chlorine atom gains an electron to form a chloride ion (Cl–).

quite different from one another, all of the group IIIA elements (except boron) have a tendency to lose three electrons to form a triply charged positive ion. An ion is a charged particle. It is formed when an atom gains or loses electrons from the shells that surround the nucleus. Atoms are usually electrically neutral, which means they carry no charge. They carry no charge because they have an equal number of positively charged protons and negatively charged electrons. However, if an atom picks up extra negatively charged electrons, it becomes a negatively charged ion (called an anion). In the same manner, if an atom loses electrons, it becomes a positively charged ion (called a cation).

Chapter Three
The Properties of Aluminum

All elements have characteristic physical and chemical properties. These properties help scientists to identify and classify them. Some examples of physical properties are an element's phase at room temperature, density, and hardness. The chemical properties of an element describe the element's ability to undergo chemical change. A chemical change converts one kind of matter into a new kind of matter. If an element undergoes chemical change easily, it is said to be very reactive.

Aluminum's Phase at Room Temperature

At room temperature, an element is found in one of three phases: solid, liquid, or gas. Knowing the phase, or physical state, of an element at room temperature helps scientists to identify it. Aluminum is found in the solid phase at room temperature. All metals are solids at room temperature, except for mercury (Hg), which is a liquid. A solid has a fixed shape and volume. Solids also resist being compressed and having their shape changed.

Aluminum's Melting Point

If solid aluminum is exposed to a high enough temperature, it will turn to liquid. The temperature at which this phase change occurs is

When aluminum is exposed to copper (II) chloride (CuCl$_2$), it dissolves. Here, some aluminum foil is placed in a beaker containing copper (II) chloride. The aluminum metal gives three electrons to copper. Copper metal is formed and falls to the bottom of the beaker as a reddish brown precipitate. At the same time, the aluminum metal turns into ions. The aluminum ions are not in solid form but dissolve into the liquid.

aluminum's melting point. For a solid to melt, the forces holding its atoms together must be overcome. Metals often have high melting points due to the strong bonds that hold their atoms together. Even so, there is a considerable variability in the melting points of metals. Mercury, for

One of the ways we can get usable aluminum is by recycling used aluminum. In order for aluminum to be reused, it must be melted. Furnaces like this one that are used to melt scrap aluminum must be very hot. Aluminum melts at about 1,221°F (660°C). Workers using the furnace must wear protective clothes and masks to shield themselves from the intense heat.

example, melts at about −38°Fahrenheit (−39°Celsius), while tungsten (W) melts at about 6,170°F (3,410°C), the highest melting point of any metallic element. Aluminum's melting point lies somewhere in between, at 1,221°F (660°C).

Aluminum's Density

Density is another physical property of matter, and each element has a unique density. Density measures how compact an object is—in other words, how much mass it contains per unit of volume. Solids often have higher densities than liquids, which, in turn, have higher densities than gases. Aluminum has a density of 2.7 g/cm^3.

In chemistry, the densities of many substances are compared to the density of water (H_2O), or 1.0 g/cm^3. If an object with a lower density than water is placed in water, it will float. However, if an object has a higher density than water, it will sink. Because aluminum has a higher density than water, pieces of aluminum will sink when dropped in water. Metals generally have high densities. Only sodium (Na), potassium (K), and lithium (Li) have densities lower than that of water.

Though aluminum has a higher density than water, it is one of the lightest of all metals. It weighs about one-third as much as an equal amount of steel. Aluminum weighs about 2.7 g/cm^3, while steel weighs about 7.8 g/cm^3. This explains why aluminum has replaced steel for many uses. For example, some parts in airplanes, automobiles, and buses are now made of aluminum rather than steel because lighter vehicles use less fuel.

Aluminum's Hardness

Aluminum is a very soft metal. Although this makes the metal very easy to shape, it means that objects made from aluminum are not very strong. On Mohs' hardness scale (see next page), aluminum has a hardness of 2.75. For comparison, your fingernail has a hardness of 2.5, and a penny has a hardness of 3.5. Aluminum is harder than your fingernail and could scratch it. However, aluminum is softer than a penny and could, in turn, be scratched by it. However, by mixing aluminum with other metals, such as copper or magnesium, it can be made as hard as steel.

Conductor of Electricity and Heat

Like other metals, aluminum is able to conduct electricity and heat. In fact, aluminum is a very good conductor, meaning that electrical current and

Mohs' Scale

Mohs' scale is used to classify the hardness of minerals, metals, and other materials. This scale was published in 1822 by an Austrian mineralogist named Friedrich Mohs. Mohs got the idea for the scale from observing miners, who routinely performed scratch tests.

The scale shows ten levels of minerals, in order of increasing hardness. In parentheses, next to the minerals that Mohs included in his chart, are some materials that you might be more familiar with. Each successive mineral is able to scratch the preceding ones and can be scratched by all that follow it.

Hardness Rating Examples

1	Talc
2	Gypsum (rock salt, fingernail)
3	Calcite (copper)
4	Fluorite (iron [Fe])
5	Apatite (cobalt [Co])
6	Orthoclase (rhodium [Rh], silicon, tungsten)
7	Quartz
8	Topaz (chromium [Cr], steel)
9	Corundum (sapphire)
10	Diamond

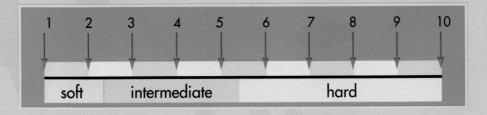

heat are able to move through aluminum easily. Metals are able to conduct electricity because the electrons from the outer shells of the atoms are able to move about from atom to atom in what is called a sea of electrons. As these electrons move, they carry the charge, and thus electricity, with them. Aluminum, along with copper and silver, is one of only three metals that are used to make electrical conduction wires. Though aluminum

Because aluminum is such a good conductor of heat, it is used to make cooking pots and utensils. However, there is some concern about aluminum from cookware leaching into foods. Some fear that consumption of aluminum could be linked to diseases like Alzheimer's, which causes memory loss. Aluminum leaching into foods is mostly only a problem when cooking highly acidic foods, such as tomato sauce. Stainless steel saucepans have largely replaced aluminum ones because of aluminum's unknown effects on the body.

Making a Fire with Aluminum

Did you know that you can make a fire with an aluminum can and a bar of chocolate? It is actually quite simple. The bottom of the can is ideal for focusing and reflecting the sun's light and energy.

However, if you look at the bottom of an aluminum can, you will notice that it is quite dull. That is where a chocolate bar comes in handy. The chocolate does an excellent job of polishing the aluminum. By rubbing chocolate on the bottom of the can and then polishing it with the wrapper or a piece of cloth, you can get it to shine. (Do not eat the chocolate afterward, however, because it will pick up aluminum from the can.)

Once the can is shiny, you need only to find a suitable piece of tinder and wait for a sunny day. Aim the bottom of your can at the sun and direct the reflected light at the tinder. You will have a campfire in no time.

does not conduct electricity as well as copper, its light weight makes it ideal for use in things such as overhead power cables.

The free-moving electrons in metals also make them good conductors of heat. When metals such as aluminum are heated, the electrons gain more energy. This makes them move about more quickly, distributing the heat throughout the whole metal. Aluminum is such a good conductor of heat that it was often used to make saucepans. Not only is aluminum an inexpensive choice for such purposes, but it also transmits heat efficiently and cools down very quickly.

Bauxite, the most important source of aluminum, is composed of a mixture of minerals. Minerals form through natural processes within the earth, such as volcanic eruptions. Magma from volcanoes, such as the Kilauea volcano in Hawaii shown here, contain elements such as iron, magnesium, and calcium. When magma cools, it crystallizes and solidifies into rock. Bauxite forms by the rapid weathering of such rocks. It is then mined and processed into aluminum metal. Heat-resistant suits, such as the one worn by this person, are often made of aluminum.

Reflector of Light and Heat

One of aluminum's most useful properties is its ability to reflect light and heat. This means that much of the light and heat that strike the surface of aluminum bounces off.

Aluminum reflects about 80 percent of the light that hits it. In fact, many of the mirrors you see around you contain aluminum. Most mirrors are made up of three layers: a protective bottom layer, a middle layer of metal (such as aluminum, silver, or tin), and a glass top layer. Because of aluminum's reflective property, the metal is also widely used as a reflector in light fixtures.

Aluminum also reflects nearly nine-tenths of the heat that reaches it. For this reason, aluminum is often used in housing insulation and as roofing material. Aluminum can be used to reflect heat back inside the house to keep it warm, or it can be used to reflect heat away from the house to keep it cool. Since aluminum reflects heat so well, it is also used to make the suits worn by firefighters. These special aluminum-coated suits reflect heat, helping to keep the firefighters safe as they walk through flames. Astronauts also wear suits coated with aluminum. These space suits help to regulate the body temperatures of the astronauts by preventing extreme heat gain or loss.

Aluminum's Reactivity

Aluminum reacts with other elements, especially oxygen, very easily. As mentioned earlier, aluminum has only three electrons in the outermost shells of its atoms. Since this outermost shell can potentially hold eight electrons, aluminum is said to have an incompletely filled shell. When atoms have incompletely filled shells, they will generally either give up those electrons to other atoms, forming what is called an ionic bond, or join together with other atoms and share them, forming what is called a covalent bond.

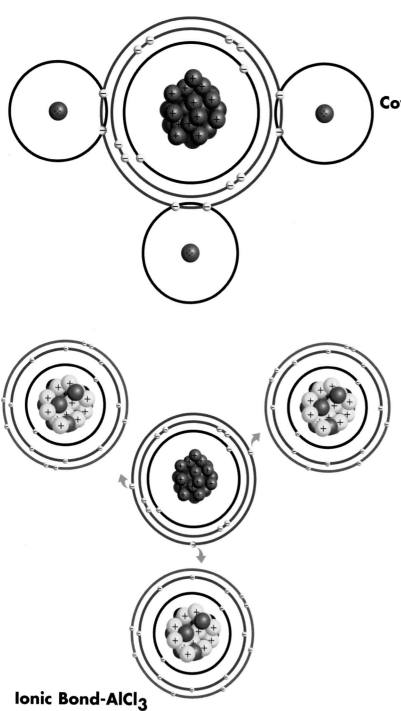

Covalent Bond-AlH₃

Ionic Bond-AlCl₃

One aluminum atom forms covalent bonds with three hydrogen atoms, forming aluminum hydride (AlH_3) *(top)*. In a covalent bond, atoms share electrons in an attempt to fill their outer shells and become more stable. The atoms are held together by a mutual attraction between the protons in their nuclei and these shared electrons. When atoms form ionic bonds, electrons are exchanged. Here, aluminum forms ionic bonds with chlorine to make aluminum trichloride ($AlCl_3$) *(bottom)*. Each chlorine atom gets three electrons to fill its shell and aluminum loses three electrons, giving it a filled outer shell also.

When elements form bonds with one another, they create compounds. Because aluminum is so reactive, in nature it is always found in compounds. This explains why it took so long for scientists to discover pure aluminum.

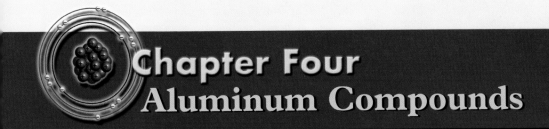
A compound is formed when two or more elements are bonded together. There are millions of different compounds all around you. When elements join together to form compounds, they lose their individual traits. Though aluminum alone is very reactive, compounds containing it can be quite stable. The traits of a compound are often very different from the traits of the individual elements from which it is made.

Aluminum is very soft, but it can be made extremely hard by bonding it with elements such as copper, magnesium, or zinc (Zn). Aluminum itself is very light, but it can be made even lighter if it is bonded to the lightest metal, lithium. Some important aluminum compounds are the oxides and the sulfates.

Aluminum Oxide

Aluminum oxide (Al_2O_3), or alumina, is probably the most important compound of aluminum. When exposed to air, aluminum quickly reacts with the oxygen in the air and becomes coated with a thin film of aluminum oxide. The aluminum oxide is a transparent film that adheres very tightly to the surface of the aluminum, making it very difficult for oxygen to react with the metal beneath. For this reason, aluminum is usually considered to

Beverages packaged in aluminum cans are sold by the millions around the world. Aluminum beverage cans were first introduced in 1965. By 1985, aluminum containers dominated the beverage market, replacing the more traditional tin-plated steel cans. Not only are aluminum cans ductile (easy to mold), light in weight, and resistant to corrosion, but they can be easily recycled.

be corrosion-resistant. Aluminum cans, for example, do not rust the way steel cans do.

Most alumina is used to make aluminum metal, but a large amount is also used for other purposes. Alumina can form corundum, one of the hardest materials in the world. Its crystals are so hard that they are often used as an abrasive in sandpaper and for various grinding tools. Because of its high melting point, it is often used to make firebricks that line the inside of ovens and furnaces. It is also used in the cosmetics industry in lotions and creams. Large crystals of corundum that contain traces of other metals are valued as gems. For example, ruby is aluminum oxide

Hydrochloric acid (HCl) dissolves many metals, including aluminum. Iron reacts rather slowly when exposed to the acid, but aluminum reacts quite rapidly. Here, some aluminum foil is being dissolved in HCl. The aluminum dissolves quite rapidly. As chlorine compounds form, hydrogen gas escapes. Aluminum's reaction with HCl explains why acidic fruit juices are never stored in aluminum containers.

that contains a small amount of chromium. Sapphire is aluminum oxide that contains trace amounts of iron and titanium (Ti). Though crystals of pure aluminum oxide are colorless, these other trace metals give the gems various beautiful colors.

Aluminum Sulfate

Aluminum sulfate ($Al_2[SO_4]_3$) is a compound of aluminum that is produced in very large quantities. Aluminum sulfate is sometimes called pickle alum due to its use in giving sourness to pickles. A lot of aluminum sulfate is used in the paper industry. When making printing paper, various materials, such as clay and rosin (a tree resin), are added to improve the paper's ability to hold ink. Aluminum sulfate is needed in order to attach the clay and rosin to the paper fibers.

Aluminum sulfate is also used to treat wastewater. It changes the surface characteristics of suspended solids, or substances floating in the water, so that they will attach to one another. Antiperspirants also contain aluminum sulfate, which acts as an astringent, or a substance used to close the openings of sweat glands.

Aluminum Putting Out Fires

Aluminum sulfate, when combined with the compound sodium bicarbonate ($NaHCO_3$), can be a source of carbon dioxide (CO_2) and foam. Carbon dioxide can be used to put out fires. For this reason, this combination was used for years to make fire extinguishers. The compounds were kept in separate compartments of the extinguisher. When the fire extinguisher was needed, the seal separating the compartments was broken, allowing them to mix. The result was aluminum hydroxide ($Al_2O_3 \cdot 3H_2O$) and carbon dioxide gas. The gas had trouble escaping through the sticky liquid. Therefore, it bubbled through it, creating a foam. This foam blanketed the fire with material that could not burn. This kept oxygen from feeding the flames, thus putting out the fire.

This helicopter is spraying foam on a forest fire in an attempt to smother it. The foam is most likely a mixture similar to that of dish soap. In order to turn the soap mixture into foam, this helicopter is equipped with the FireSno foam delivery system. The mixture is shot from pressurized tanks. As the liquid mixture slams into the air, it turns into foam. The reaction of some aluminum compounds with sodium bicarbonate can also produce foam resulting from gases being trapped in its liquid product.

Aluminum Hydroxide

When aluminum oxide is hydrated (or containing water) it is known as aluminum hydroxide ($Al_2O_3 \bullet 3H_2O$). This compound is a white, jellylike substance that is formed when an alkali, or base, is added to an aluminum salt. Toothpaste often contains small amounts of aluminum hydroxide. It helps to counteract the buildup of acids in the mouth that destroy the enamel layer of our teeth.

Aluminum Chloride

Aluminum chloride ($AlCl_3$) is a white solid. It is made in a laboratory by passing dry chlorine (Cl_2) gas or dry hydrogen chloride (HCl) gas over a heated sample of aluminum. Aluminum chloride reacts readily with water and turns to fumes when in contact with moist air. Like aluminum sulfate, aluminum chloride is an astringent and is therefore an ingredient in many antiperspirants.

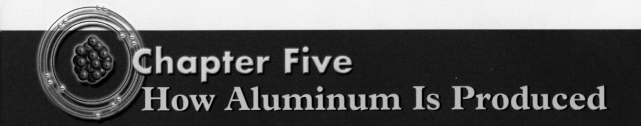

A total of about twenty-nine million tons of aluminum is needed each year to meet worldwide demand for the metal. Aluminum metal is made from two sources: primary metal, which is produced from ore, and secondary metal, which is produced from scrap. About twenty-two million tons of the aluminum that is made comes from ore, while seven million tons of it comes from recycled material. The United States is the largest consumer and the fourth-largest producer of the metal in the world.

Aluminum from Ore

Most of the aluminum in the world today is made from bauxite. Bauxite is composed mostly of aluminum hydroxide, meaning that it is made of alumina and water. Since bauxite is found close to the surface of the ground, mining it is fairly easy. Bulldozers expose the bauxite deposits by clearing away vegetation and topsoil. The bauxite is then broken up with explosives and gathered into trucks and trains by powerful mechanical diggers.

After the bauxite has been mined, it is sent to a processing plant where the water is removed, leaving behind alumina. Australia, the United States, and China are the world's largest producers of alumina. Because it does not contain water, alumina is much lighter than bauxite. It flows easily through the processing plants, unlike bauxite, which has a

Metallic aluminum is produced from alumina, a white, powdery compound that contains aluminum. The process starts when bauxite, an aluminum ore, is mined from the ground. The bauxite is then processed into alumina using the Bayer process. Many countries around the world mine bauxite and use it to produce alumina. The alumina shown in this picture was produced at a factory in Pavlodar, Kazakhstan. It was then sold to aluminum smelters who turned it into aluminum metal.

sticky, muddy consistency. The inexpensive process developed to extract the alumina from bauxite is called the Bayer process. It is named after Austrian chemist Karl Joseph Bayer, who pioneered the process in 1888.

The Bayer Process

The Bayer process refines bauxite into almost pure alumina by mixing the bauxite with sodium hydroxide (NaOH) and water at a high temperature and pressure. This results in a boiling hot solution of sodium aluminate ($NaAl[OH]_4$). This solution is drained into tanks, where impurities are filtered out. The resulting liquid is then cooled in cooling vats.

As the liquid cools, aluminum hydroxide crystals form. The crystals are washed and heat-dried in ovens at temperatures of over 1760°F (960°C) in order to drive off any remaining moisture. The resulting white, granular powder is alumina. This alumina is sent to a refinery, where pure aluminum will be produced from it.

The Hall-Héroult Process

Aluminum metal is refined from alumina in a process called the Hall-Héroult process, named after Charles Martin Hall and French metallurgist Paul-Louis-Toussaint Héroult, who discovered the process independently of one another in 1886. In the Hall-Héroult process, alumina is dissolved in molten cryolite (Na_3AlF_6). Cryolite is another aluminum-containing mineral. In addition to aluminum, cryolite contains sodium and fluorine (F). At one time, cryolite was found in large quantities on the west coast of Greenland. However, that supply ran out in 1987. Though small quantities can still be found in various locations around the world, the use of cryolite has mostly been replaced by artificially produced sodium aluminum fluoride.

Dissolving the alumina in the molten bath allows it to dissociate, or break down into its ions, Al^{3+} and O^{2-}. Electricity is passed through the molten bath in a process called electrolysis. Since the oxygen ions are negatively charged, they move through the solution to the positively charged electrode of the bath, or the anode. Here, free oxygen, O_2, is released.

Meanwhile, the positively charged aluminum ions are drawn toward the negatively charged electrode of the bath, or cathode, which lines the bath. This causes a layer of molten aluminum to form at the bottom of the

Without recycling, piles of aluminum scrap, like this one, would sit in land-fills for hundreds of years because of aluminum's resistance to corrosion. Though more than fifty percent of America's aluminum is recycled, about two million tons of used aluminum are thrown away each year.

bath. For about every four tons of bauxite used in these processes, a ton of aluminum will be produced.

Obtaining aluminum through recycling instead of through the refinement of aluminum ore saves a considerable amount of energy and money. Because scrap aluminum has already been refined, all of the energy needed for the Bayer and Hall-Héroult processes is saved when aluminum is recycled. Aluminum scrap needs only to be melted down in order to be reused. Therefore, recycling uses about one-twentieth of the energy it took to produce the aluminum in the first place. Fewer resources, such as coal and oil, are used, and fewer pollutants, such as carbon dioxide and sulfur dioxide (SO_2), are released. Recycling also eliminates the cost of mining and shipping.

Chapter Six
Aluminum and You

A rocket is a device that carries objects through air and space. Today, rockets carry explosive devices to targets, boost satellites and other spacecraft into space, and carry scientific instruments into the upper atmosphere. Rockets are propelled by both fuel and an oxidizer, or an oxygen-containing substance.

There are two types of rocket fuel: liquid propellants and solid propellants. Liquid propellants are usually composed of liquid hydrogen and liquid oxygen. Solid rocket propellants contain aluminum metal powder and an oxidizer of ammonium perchlorate (NH_4ClO_4). Without the solid rocket boosters that use this fuel, shuttles would not be able to make it out of orbit. In fact, almost three-fourths of a shuttle's total rocket power at launch comes from these reusable boosters. When a solid rocket booster is launched into the sky, a cloud of aluminum oxide billows out like baby powder as the rocket is burned.

Aluminum and Health

Compounds that contain aluminum are sometimes found in the human body. Though there is no evidence that aluminum is harmful to us in small doses, physicians believe that a buildup of the metal in the body may cause health problems. Aluminum can enter the body in many ways.

Some antacids, or drugs used to reduce the amount of acid in the stomach, contain aluminum. Aluminum may also enter foods prepared with aluminum pots, utensils, or foil. Some food additives, such as potassium alum ($KAl[SO_4]_2$), used to whiten flour, also contain aluminum. As we have seen, aluminum chloride and aluminum sulfate are used to make antiperspirants, and toothpastes contain aluminum hydroxide. Other aluminum-containing compounds are also used in cosmetics.

Though it is believed that most aluminum is excreted from the body, some can build up in the brain, thyroid gland, liver, and lungs. Some scientists believe that aluminum is linked to diseases such as emphysema, fibrosis, lung disease, and Alzheimer's disease. For instance, there is some evidence that Alzheimer's disease, which destroys a person's memory, is more common in areas where the water contains a lot of aluminum.

Aluminum and Recycling

The recycling of aluminum is very important. Though there is still a lot of aluminum in the ground to be mined, the bauxite mining has already devastated large areas of land. Aluminum has no limit to the number of times it can be recycled, yet every day one hundred million beverage cans are sent to landfills, incinerated, or thrown on the ground as litter in the United States. Since aluminum does not usually corrode, or rust, aluminum cans in dumps may remain intact for decades.

There are many ways that you can help to make sure that aluminum is being recycled. Find local recycling centers in your area. You can take your cans to a recycling center, or if your community offers curbside recycling, you can simply place your empty aluminum cans in the bins that are provided. Your cans will be picked up and taken to a recycling facility. One hundred percent of aluminum cans that are recycled end up as other aluminum products in as little as sixty days. Any action you take to contribute to recycling can make a world of difference.

Aluminum cans that are brought to recycling centers are crushed and formed into large blocks. These blocks of crushed aluminum cans are then sent to other facilities where they are shredded to reduce volume and heated to remove coatings and moisture. They are then put into a furnace where they are melted and formed into bars. The bars are rolled into sheets and then, finally, formed into new cans.

Hans Christian Ørsted had no idea how much his life-changing discovery would benefit the world when he separated aluminum from its ore. From aluminum cans to airplane parts to electrical wires, the contribution that aluminum has made to modern society is impossible to measure.

The Periodic Table of Elements

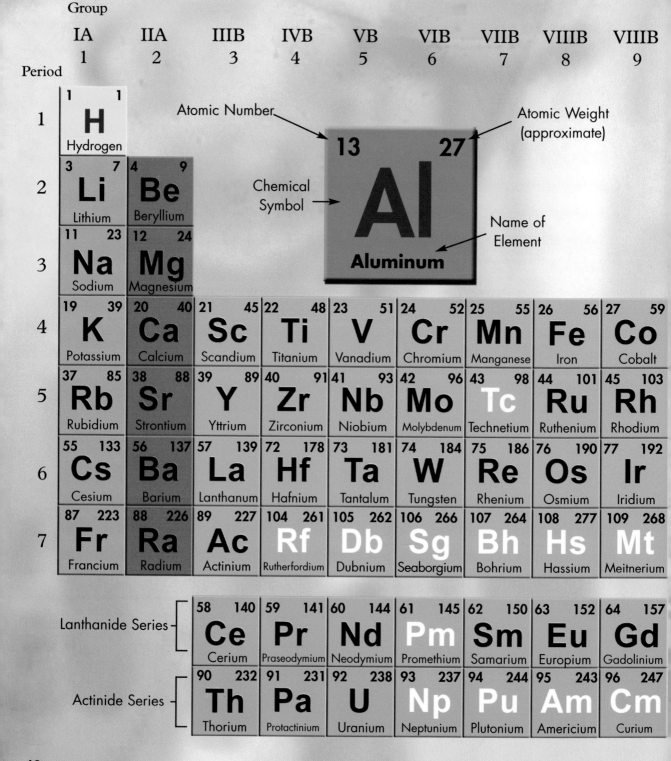

Group

IA	IIA	IIIB	IVB	VB	VIB	VIIB	VIIIB	VIIIB
1	2	3	4	5	6	7	8	9

Period

Atomic Number

Atomic Weight (approximate)

13   27

Al

Chemical Symbol

Name of Element

Aluminum

Period 1:
1 1 **H** Hydrogen

Period 2:
3 7 **Li** Lithium — 4 9 **Be** Beryllium

Period 3:
11 23 **Na** Sodium — 12 24 **Mg** Magnesium

Period 4:
19 39 **K** Potassium — 20 40 **Ca** Calcium — 21 45 **Sc** Scandium — 22 48 **Ti** Titanium — 23 51 **V** Vanadium — 24 52 **Cr** Chromium — 25 55 **Mn** Manganese — 26 56 **Fe** Iron — 27 59 **Co** Cobalt

Period 5:
37 85 **Rb** Rubidium — 38 88 **Sr** Strontium — 39 89 **Y** Yttrium — 40 91 **Zr** Zirconium — 41 93 **Nb** Niobium — 42 96 **Mo** Molybdenum — 43 98 **Tc** Technetium — 44 101 **Ru** Ruthenium — 45 103 **Rh** Rhodium

Period 6:
55 133 **Cs** Cesium — 56 137 **Ba** Barium — 57 139 **La** Lanthanum — 72 178 **Hf** Hafnium — 73 181 **Ta** Tantalum — 74 184 **W** Tungsten — 75 186 **Re** Rhenium — 76 190 **Os** Osmium — 77 192 **Ir** Iridium

Period 7:
87 223 **Fr** Francium — 88 226 **Ra** Radium — 89 227 **Ac** Actinium — 104 261 **Rf** Rutherfordium — 105 262 **Db** Dubnium — 106 266 **Sg** Seaborgium — 107 264 **Bh** Bohrium — 108 277 **Hs** Hassium — 109 268 **Mt** Meitnerium

Lanthanide Series:
58 140 **Ce** Cerium — 59 141 **Pr** Praseodymium — 60 144 **Nd** Neodymium — 61 145 **Pm** Promethium — 62 150 **Sm** Samarium — 63 152 **Eu** Europium — 64 157 **Gd** Gadolinium

Actinide Series:
90 232 **Th** Thorium — 91 231 **Pa** Protactinium — 92 238 **U** Uranium — 93 237 **Np** Neptunium — 94 244 **Pu** Plutonium — 95 243 **Am** Americium — 96 247 **Cm** Curium

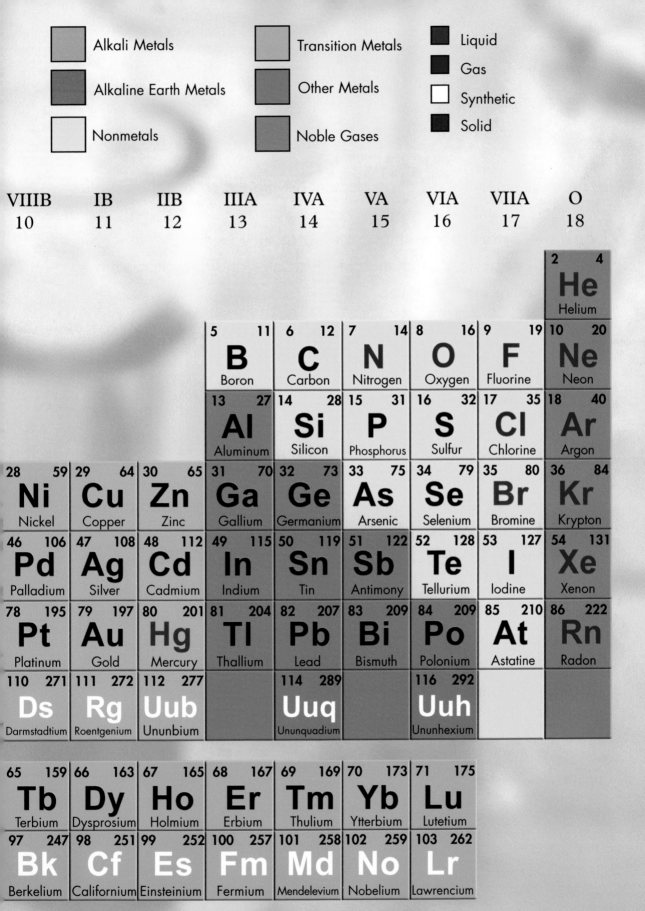

Legend

- Alkali Metals
- Alkaline Earth Metals
- Nonmetals
- Transition Metals
- Other Metals
- Noble Gases
- Liquid
- Gas
- Synthetic
- Solid

VIIIB 10	IB 11	IIB 12	IIIA 13	IVA 14	VA 15	VIA 16	VIIA 17	O 18
								2 4 **He** Helium
			5 11 **B** Boron	6 12 **C** Carbon	7 14 **N** Nitrogen	8 16 **O** Oxygen	9 19 **F** Fluorine	10 20 **Ne** Neon
			13 27 **Al** Aluminum	14 28 **Si** Silicon	15 31 **P** Phosphorus	16 32 **S** Sulfur	17 35 **Cl** Chlorine	18 40 **Ar** Argon
28 59 **Ni** Nickel	29 64 **Cu** Copper	30 65 **Zn** Zinc	31 70 **Ga** Gallium	32 73 **Ge** Germanium	33 75 **As** Arsenic	34 79 **Se** Selenium	35 80 **Br** Bromine	36 84 **Kr** Krypton
46 106 **Pd** Palladium	47 108 **Ag** Silver	48 112 **Cd** Cadmium	49 115 **In** Indium	50 119 **Sn** Tin	51 122 **Sb** Antimony	52 128 **Te** Tellurium	53 127 **I** Iodine	54 131 **Xe** Xenon
78 195 **Pt** Platinum	79 197 **Au** Gold	80 201 **Hg** Mercury	81 204 **Tl** Thallium	82 207 **Pb** Lead	83 209 **Bi** Bismuth	84 209 **Po** Polonium	85 210 **At** Astatine	86 222 **Rn** Radon
110 271 **Ds** Darmstadtium	111 272 **Rg** Roentgenium	112 277 **Uub** Ununbium		114 289 **Uuq** Ununquadium		116 292 **Uuh** Ununhexium		

65 159 **Tb** Terbium	66 163 **Dy** Dysprosium	67 165 **Ho** Holmium	68 167 **Er** Erbium	69 169 **Tm** Thulium	70 173 **Yb** Ytterbium	71 175 **Lu** Lutetium
97 247 **Bk** Berkelium	98 251 **Cf** Californium	99 252 **Es** Einsteinium	100 257 **Fm** Fermium	101 258 **Md** Mendelevium	102 259 **No** Nobelium	103 262 **Lr** Lawrencium

Glossary

atom The smallest part of an element having the chemical properties of that element.

bond An attractive force that links two atoms together.

chemical reaction A change in which one kind of matter is turned into another kind of matter.

crust The surface layer of the earth.

electrolysis The process in which electricity is passed through a liquid between electrodes.

mass The amount of matter an object contains.

matter Anything that takes up space and has mass.

orbitals The regions of space around a nucleus where electrons are most likely to be found.

ore A mineral deposit containing something that can be profitably mined.

oxide A compound that contains oxygen.

precipitate A solid formed in a solution as the result of a chemical reaction.

pure Referring to a material that contains a single kind of atom.

refine To purify something.

shell A group of orbitals.

volume The amount of space that something occupies.

For More Information

Aluminum Company of America
Alcoa Corporate Center
201 Isabella Street
Pittsburgh, PA 15212-5858
(412) 553-4545
Web site: http://www.alcoa.com/global/en/home.asp

Southern Aluminum Finishing Company
Corporate Headquarters
1581 Huber Street NW
Atlanta, GA 30318-3781
(800) 241-7429
Web site:http://www.saf.com/index.html

Web Sites

Due to the changing nature of Internet links, the Rosen Publishing Group, Inc., has developed an online list of Web sites related to the subject of this book. This site is updated regularly. Please use this link to access the list:

http://www.rosenlinks.com/uept/alum

Cox, P. A. *The Elements: Their Origins, Abundance, and Distribution.* New York, NY: Oxford University Press, 1989.

Greenwood, N. N., and A. Earnshaw. *Chemistry of the Elements.* Oxford, England: Pergamon Press, 1984.

Heiserman, David L. *Exploring Chemical Elements and Their Compounds.* Blue Ridge Summit, PA: Tab Books, 1992.

Hudson, John. *The History of Chemistry.* New York, NY: Routledge, 1992.

Newton, David E. *The Chemical Elements.* New York, NY: Franklin Watts, 1994.

Saunders, Nigel. *Aluminum and the Elements of Group 13.* Chicago, IL: Heinemann; 2004.

Snyder, Carl H. *The Extraordinary Chemistry of Ordinary Things.* New York, NY: Wiley, 1995.

Brady, James E., and John R. Holum. *Chemistry: The Study of Matter and Its Changes.* New York, NY: John Wiley & Sons, 1993.

Ebbing, Darrell D. *General Chemistry.* 4th ed. Boston, MA: Houghton Mifflin Company, 1993.

Farndon, John. *The Elements: Aluminum.* New York, NY: Marshall Cavendish Corporation, 2001.

Stwertka, Albert. *A Guide to the Elements.* 2nd ed. New York, NY: Oxford University Press, 2002.

About the Author

Heather Elizabeth Hasan graduated from college summa cum laude with a dual major in biochemistry and chemistry. She has written numerous books for the Rosen Publishing Group, including *Understanding the Elements of the Periodic Table: Iron* and *Understanding the Elements of the Periodic Table: Nitrogen*. She currently lives in Greencastle, Pennsylvania, with her husband, Omar, and their sons, Samuel and Matthew.

Photo Credits

Cover, pp. 1, 11, 13, 15, 16, 26, 40–41 by Tahara Anderson; p. 7 © George Whitely/Photo Researchers, Inc.; p. 8 © Boyer/Roger Viollet/ Getty Images; pp. 9, 31 © AP/Wide World Photos; pp. 18, 29 by Maura McConnell; pp. 19, 36 © Chris Knapton/Photo Researchers, Inc.; p. 22 © Elizabeth Whiting and Associates/Corbis; p. 24 © Kraft/ Photo Researchers, Inc.; p. 28 © Gary Gladstone/Corbis; p. 34 © Robert Nickelsberg/Time Life Pictures/Getty Images; p. 39 © Will and Deni McIntyre/Photo Researchers, Inc.

Special thanks to Megan Roberts, director of science, Region 9 Schools, New York City, NY, and Jenny Ingber, high school chemistry teacher, Region 9 Schools, New York City, NY, for their assistance in executing the science experiments illustrated in this book.

Designer: Tahara Anderson